NOBEL PRIZE

An in-depth view of Nobel prizes and Laureates

Carlos J. Garcia

Copyright

All rights reserved. No part of this publication may be reproduced, distributed, or transmitted in any form or by any means, including photocopying, recording, or other electronic or mechanical methods, without the prior written permission of the publisher, except in the case of brief quotations embodied in critical reviews and certain other noncommercial uses permitted by copyright law.

Copyright © Carlos J. Garcia, 2022.

Table of content

Introduction

Chapter 1: The Nobel Prize

Chapter 2: Nobel Prize Categories

Chapter 3: Award process

Chapter 4: Years without Nobel Prizes

Chapter 5: 2022 Nobel Prizes

Chapter 6: Nobel prize facts

Introduction

Nobel

Alfred Bernhard Nobel (/noʊˈbɛl/) was a Swedish chemist, engineer, inventor, businessman, and philanthropist.

He is most recognized for having left his riches to create the Nobel Prize, however, he also made numerous notable contributions to science, holding 355 patents throughout his lifetime.

Nobel's most renowned innovation was dynamite, a safer and simpler technique of harnessing the explosive potential of nitroglycerin; it was patented in 1867.

Nobel demonstrated an early talent for science and study, notably in chemistry and languages; he became proficient in six languages and submitted his first patent at age 24. He embarked on several economic endeavors with his family, most notably owning Bofors, an iron and steel factory

where he grew into a large maker of cannons and other weaponry.
The synthetic element nobelium was named after him, and his name and legacy also endure in firms such as Dynamit Nobel and AkzoNobel, which derive from mergers with enterprises he created.

In 1888, the death of his brother Ludvig forced some newspapers to print obituaries of Alfred in mistake. One French newspaper attacked him for his creation of military explosives, dynamite, (which was mostly utilized for civilian applications) and is supposed to have brought to his determination to leave a better legacy after his death.
The obituary said, Le Marchand de la mort est mort ("The merchant of death is dead"), and went on to explain, "Dr. Alfred Nobel, who grew wealthy by developing methods to murder more people quicker than ever before, died yesterday."

Nobel read the obituary and was shocked at the notion that he would be remembered in this manner.

After reading the erroneous obituary portraying him as a war profiteer, Nobel was motivated to gift his riches to the Nobel Prize organization, which would yearly prize people who "conferred the greatest value to humanity".

His choice to posthumously contribute the bulk of his riches to create the Nobel Prize has been linked at least in part to his wishing to leave behind a better legacy.

On 27 November 1895, at the Swedish-Norwegian Club in Paris, Nobel wrote his final will and put aside the majority of his wealth to create the Nobel Prize, to be granted yearly without difference of country.

After taxes and bequests to people, Nobel donated 94% of his entire assets, 31,225,000 Swedish kronor, to create the five Nobel

Prizes. This translated to £1,687,837 (GBP) at the time.

In 2012, the capital was valued at roughly SEK 3.1 billion (US$472 million, EUR 337 million), which is over double the amount of the founding capital, taking inflation into account.

The first three of these prizes are awarded for eminence in physical science, chemistry, and medical science or physiology; the fourth is for literary work "in an ideal direction" and the fifth prize is to be given to the person or society that renders the greatest service to the cause of international fraternity, in the suppression or reduction of standing armies, or the establishment or furtherance of peace congresses.

The phrasing for the literary award granted for a work "in an ideal direction", is obscure and has created considerable uncertainty. For many years, the Swedish Academy understood "ideal" as "idealistic" and used it as a pretext not to present the prize to famous but less romantic writers, such as Henrik Ibsen and Leo Tolstoy. This view has subsequently been updated, and the prize has been granted to, for example, Dario Fo and José Saramago, who do not belong to the camp of literary idealism.

There was an opportunity for interpretation by the bodies he had selected for deciding on the physical sciences and chemistry awards, given that he had not contacted them before writing the will. In his one-page bequest, he specified that the money goes to discoveries or innovations in the physical sciences and discoveries or advancements in chemistry. He had opened the door to technical honors but had not provided guidelines on how to cope with the divide between science and technology.

Since the determining bodies he had picked were more concerned with the former, the prizes went to scientists more frequently than engineers, technicians, or other inventions.

Sweden's central bank Sveriges Riksbank celebrated its 300th anniversary in 1968 by giving a considerable quantity of money to the Nobel Foundation to be used to establish the sixth award in the area of economics in memory of Alfred Nobel.
In 2001, Alfred Nobel's great-great-nephew, Peter Nobel (born 1931), petitioned the Bank of Sweden to separate its prize to economists presented "in Alfred Nobel's memory" from the five other awards. This request adds to the issue of whether the Bank of Sweden Prize in Economic Sciences in Memory of Alfred Nobel is indeed a valid "Nobel Prize".

Nobel was chosen as a member of the Royal Swedish Academy of Sciences, which, due to his wishes, would be responsible for picking the Nobel laureates in physics and chemistry.

Chapter 1: The Nobel Prize

The Nobel prizes are according to Alfred Nobel's will of 1895, are presented to "those who, during the previous year, have bestowed the greatest benefit to humanity.". The Nobel prize is recognized as the most prestigious prize in the world and are presented yearly by four institutions (three Swedish and one Norwegian) from a fund created under the will of Alfred B. Nobel.

Each Nobel Prize consists of a gold medal, a certificate containing a citation, and a quantity of money, the amount of which depends on the revenue of the Nobel Foundation.

Each recipient (known as a "laureate") receives;

   1) a gold medal

Each medal displays a picture of Alfred Nobel in the left profile on the obverse. The medals for physics, chemistry, physiology or medicine, and literature have similar obverses, depicting the picture of Alfred Nobel and the years of his birth and death. Nobel's face also appears on the obverse of the Peace Prize medal and the medal for the Economics Prize, albeit with a slightly different design. For instance, the laureate's name is inscribed on the rim of the Economics medal. The picture on the back of a medal changes according to the institution granting the honor. The back sides of the medals for chemistry and physics have the same design.

All medals manufactured before 1980 were minted in 23-carat gold. Since then, they have been struck in 18-carat green gold coated with 24-carat gold. The weight of each medal varies with the value of gold but averages roughly 175 grams (0.386 lb) for each medal.

The diameter is 66 millimeters (2.6 in) and the thickness ranges between 5.2 millimeters (0.20 in) and 2.4 millimeters (0.094 in) (0.094 in). Because of the high value of its gold content and inclination to be on public display, Nobel awards are prone to medal theft.

2) a diploma
Nobel laureates receive a diploma straight from the hands of the King of Sweden, or in the case of the peace prize, the head of the Norwegian Nobel Committee. Each certificate is specifically prepared by the prize-awarding organizations for the laureates who receive it.
The diploma comprises a picture and text in Swedish which indicates the name of the recipient and generally a citation of why they earned the award. None of the Nobel Peace Prize laureates has ever had a citation on their diplomas.

3) a monetary prize
The laureates are given a sum of money when they collect their awards, in the form of a paper certifying the amount received. The amount of prize money depends upon how much money the Nobel Foundation can

pay each year. The purse expanded during the 1980s when the prize money was 880,000 SEK for every award (c. 2.6 million SEK overall, US$350,000 now). In 2009, the monetary prize was 10 million SEK (US$1.4 million). In June 2012, it dropped to 8 million SEK.

If two laureates share the prize in a category, the award grant is shared evenly between the winners. If there are three, the awarding committee has the option of splitting the grant evenly or granting one-half to one recipient and one-quarter to each of the others. It is typical for awardees to contribute prize money to promote scientific, cultural, or humanitarian purposes.

A Nobel Prize is either awarded totally to one person, split evenly between two individuals, or shared by three persons. In the second situation, each of the three individuals may earn a one-third part of the prize, or two together can receive a one-half share, however, the Nobel Peace Prize can be granted to groups comprising more than three people.

Sometimes a prize is retained until the next year; if not then granted, it is put back into the coffers, which occurs likewise when a prize is neither awarded nor reserved. Two awards in the same field (i.e., the prize withheld from the previous year and the current year's prize can therefore be granted in one year).
If a prize is denied or not accepted before a specific deadline, the prize money goes back into the treasury. Some awards have been rejected by their recipients, and in select situations, governments have refused to allow their people to take them. Those who receive a prize are nonetheless placed into the list of Nobel laureates with the phrase "declined the award." Motives for nonacceptance may vary, but most often the reason has been external pressure; for example, in 1937 Adolf Hitler forbade

Germans in the future from accepting Nobel Prizes because he had been infuriated by the award of the 1935 Peace Prize to the anti-Nazi journalist Carl von Ossietzky, who at the time was a political prisoner in Germany.

In certain instances, the refuser eventually revealed the actual reason behind the denial and was handed the Nobel gold medal and the diploma—but not the money, which generally reverts to the funds after a specified amount of time.

Although Nobel Prizes are not given posthumously, if a person is granted a prize and dies before receiving it, the prize is presented.

Chapter 2: Nobel Prize Categories

Since 1901, Nobel Prizes have been awarded in the fields of Physics, Chemistry, Physiology or Medicine, Literature, and Peace (Nobel characterized the Peace Prize as "to the person who has done the most or best to advance fellowship among nations, the abolition or reduction of standing armies, and the establishment and promotion of peace congresses").

In 1968, Sveriges Riksbank (Sweden's central bank) funded the establishment of the Prize in Economic Sciences in Memory of Alfred Nobel, to also be administered by the Nobel Foundation.

Prizes are withheld when no worthy candidate can be found or when the world situation prevents the gathering of information required to reach a decision, as happened during World Wars I and II.

The prizes are open to all, irrespective of nationality, race, creed, or ideology. They can be awarded more than once to the same recipient.

The ceremonial presentations of the awards for physics, chemistry, physiology or medicine, literature, and economics take place in Stockholm, and that for peace takes place in Oslo on the anniversary of Nobel's death(December 10).

The laureates usually receive their prizes in person, and each presents a lecture in connection with the award ceremonies.

The general principles governing awards were laid down by Alfred Nobel in his will.

In 1900 extra norms of interpretation and administration were agreed upon between the executors, representatives of the prize-awarding organizations, and the Nobel family and were endorsed by the monarch in the council. These legislative regulations have overall remained constant but have been slightly modified in implementation.

For example, Nobel's rule that the awards be granted for accomplishments accomplished during "the previous year" was manifestly untenable for scientists and even authors, the actual value of whose discoveries, studies, or works would not be broadly evident for many years.

Nobel's vague provision that the literary prize is granted to the writers of works with an "idealistic tendency" was construed rigidly in the beginning but has subsequently been interpreted more freely. The premise for the economics prize was scientific—i.e., mathematical or statistical, rather than political or social.

The Nobel Prizes for physics, chemistry, and physiology or medicine have traditionally been the least contentious, whereas those for literature and peace have been, by their very nature, the most vulnerable to critical criticism.

The Peace Award has been the prize most often reserved or withheld.

Below are some information about the Nobel Prize categories and Laureates;

Total number of physics prizes awarded: 116
Total number of Physics laureates: 222

Total number of Chemistry prizes awarded: 114
Total number of Chemistry laureates: 191

Total number of medicine prizes awarded: 113
Total number of medicine laureates: 225

Total number of Literature prizes awarded: 115
Total number of Literature laureates: 119

Total number of Peace prizes awarded: 103
Total number of Peace laureates: 140

Total number of Economic sciences prizes awarded: 54

Total number of Economic sciences laureates: 92

Chapter 3: Award process

The award procedure is identical for all of the Nobel Prizes, the key distinction being who may submit nominations for each of them.

Nominations
Nomination forms are issued by the Nobel Committee to roughly 3,000 persons, generally, in September the year before the awards are presented. These persons are often notable academics working on a related topic. Regarding the Peace Prize, questions are also directed to governments, previous Peace Prize laureates, and current or former members of the Norwegian Nobel Committee.

The deadline for the return of the nomination forms is 31 January of the year of the award. The Nobel Committee nominates roughly 300 possible laureates from these forms and other names.

The candidates are not publicly identified, nor are they aware that they are being considered for the honor. All nomination records for a prize are sealed for 50 years after the granting of the prize.

Selection
The Nobel Committee then develops a report reflecting the recommendations of experts in the relevant subjects. This, along with the list of preliminary candidates, is sent to the prize-awarding organizations. There are four awarding institutions for the six awards awarded:
Royal Swedish Academy of Sciences.
Nobel Assembly at the Karolinska Institute.
Swedish Academy.
Norwegian Nobel Committee.

The institutions gather to determine the laureate or laureates in each discipline by a majority vote. Their decision, which cannot be disputed, is announced immediately after the voting.

A maximum of three laureates and two separate works may be chosen for every prize. Except for the Peace Prize, which may be presented to organizations, the prizes can only be given to people.

Posthumous nominations
Although posthumous nominations are not now accepted, persons who died in the months between their nomination and the judgment of the award committee were previously eligible to win the prize. This has happened twice: the 1931 Literature Prize was granted to Erik Axel Karlfeldt, and the 1961 Peace Prize was presented to UN Secretary-General Dag Hammarskjöld.
Since 1974, laureates must be deemed alive at the time of the October announcement. There has been one recipient, William Vickrey, who in 1996 died after the award (in Economics) was announced but before it could be delivered.

On 3rd October 2011, the laureates for the Nobel Prize in Physiology or Medicine were revealed; however, the committee was not aware that one of the laureates, Ralph M. Steinman, had died three days before.
The committee was discussing Steinman's prize, as the norm stipulates that the prize is not granted posthumously. The committee eventually ruled that while the decision to give Steinman the prize "was taken in good faith", it would stay intact.

Nobel's will called for awards to be granted in recognition of discoveries made "during the prior year". Early on, the prizes mainly highlighted fresh findings.
However, several of those early findings were eventually dismissed. For example, Johannes Fibiger was given the 1926 Prize in Physiology or Medicine for his supposed discovery of a parasite that caused cancer. To prevent repeating this humiliation, the awards increasingly prizeed scientific

breakthroughs that have survived the test of time.

According to Ralf Pettersson, past head of the Nobel Prize Committee for Physiology or Medicine, "the condition 'the previous year' is read by the Nobel Assembly as the year when the entire significance of the discovery has become clear."

The delay between the award and the achievement it celebrates varies from discipline to discipline.

The Literature Prize is often granted to commemorate a cumulative lifetime body of work rather than a single accomplishment.

The Peace Prize may also be given for a lifelong corpus of effort. For example, 2008 laureate Martti Ahtisaari was prizeed for his efforts to address international problems. However, they may also be granted for particular recent occurrences. For instance, Kofi Annan was given the 2001 Peace Prize barely four years after becoming the Secretary-General of the United Nations.

Similarly, Yasser Arafat, Yitzhak Rabin, and Shimon Peres got the 1994 medal, roughly a year after they successfully finalized the Oslo Accords.

The most recent controversy was sparked by giving the 2009 Nobel Peace Prize to Barack Obama in his first year as US President.

Awards for physics, chemistry, and medicine are normally granted after the accomplishment has been widely acknowledged. Sometimes, this takes decades - for example, Subrahmanyan Chandrasekhar shared the 1983 Physics Prize for his 1930s work on star structure and development. Not all scientists live long enough for their work to be acknowledged. Some discoveries can never be considered for a prize if their importance is discovered after the discoverers have died.

In Summary, the award process proceeds thus;

The reputation of the Nobel Prize originates in part from the significant study that goes into the selection of the prizewinners. Although the winners are revealed in October and November, the selection process starts in the early fall of the prior year, when the prize-awarding organizations seek more than 6,000 persons to submit or nominate candidates for the awards. Some 1,000 individuals submit nominations for each award, and the number of candidates normally fluctuates from 100 to roughly 250. Among those nominating are Nobel laureates, members of the prize-awarding organizations themselves; scientists engaged in the disciplines of physics, chemistry, economics, and physiology or medicine; and administrators and members of numerous universities and learned academies. The replies must submit a written proposal that outlines their candidates' merits. Self-nomination immediately disqualifies the candidate. Prize suggestions must be submitted to the Nobel Committees on or by January 31 of the award year.

On February 1 the six Nobel Committees (one for each award category) start their work on the nominations submitted. Outside experts are frequently consulted during the process to help the committees determine the originality and significance of each nominee's contribution. During September and early October, the Nobel Committees accomplished their work and submit their recommendations to the Royal Swedish Academy of Sciences and other prize-awarding institutions. A committee's recommendation is usually but not invariably followed. The debates and the vote within these entities remain confidential at all levels. The final judgment of the awardees must be made by November 15.

Prizes may be presented solely to people, save the Peace Prize, which may also be bestowed upon an organization. A person may not be nominated posthumously, however, a winner who dies before receiving the prize may be given it posthumously, as with Dag Hammarskjöld (for peace; 1961), Erik Axel Karlfeldt (for literature; 1931), and Ralph M. Steinman (for physiology or medicine; 2011). (for physiology or medicine; 2011). (Steinman was awarded a winner many days after his death, which was undisclosed to the Nobel Assembly. It was ruled that he would remain a Nobel laureate, as the aim of the posthumous rule was to prevent awards being intentionally given to dead persons.) The awards may not be challenged. Official support, whether diplomatic or political, for a given candidate, has no influence on the award process since the prize awarders, as such, are independent of the state.

Chapter 4: Years without Nobel Prizes

Since the outset, in 1901, there are several years when the Nobel Prizes have not been given. The total number of times is 49. Most of them during World War I (1914-1918) and II (1939-1945).

The laws of the Nobel Foundation says: "If none of the works under consideration is deemed to be of the significance mentioned in the first paragraph, the prize money must be held until the following year. If even then, the prize cannot be granted, the sum must be put to the Foundation's restricted funds."

Below are years when Nobel prizes were not awarded;

Physics: 1916, 1931, 1934, 1940, 1941, 1942

Chemistry: 1916, 1917, 1919, 1924, 1933, 1940, 1941, 1942

Physiology or medicine: 1915, 1916, 1917, 1918, 1921, 1925, 1940, 1941, 1942

Literature: 1914, 1918, 1935, 1940, 1941, 1942, 1943

Peace: 1914, 1915, 1916, 1918, 1923, 1924, 1928, 1932, 1939, 1940, 1941, 1942, 1943, 1948, 1955, 1956, 1966, 1967, 1972

Economic sciences: None

Chapter 5: 2022 Nobel Prizes

The Nobel Prize in Physics 2022
The Nobel Prize in Physics is given by the Royal Swedish Academy of Sciences, Stockholm, Sweden.
The Nobel Prize in Physics 2022 was given to Alain Aspect, John F. Clauser, and Anton Zeilinger "for work with entangled photons, showing the violation of Bell inequalities and pioneering quantum information science".

Using pioneering experiments, Alain Aspect, John Clauser, and Anton Zeilinger have proven the possibility to analyze and manipulate particles that are in entangled states.
What happens to one particle in an entangled pair dictates what happens to the other, even though they are truly too far away to impact one other.
The laureates' invention of experimental instruments has created the groundwork for

a new age of quantum technology, and also paved the door for new technologies based upon quantum information.

The Nobel Prize in Chemistry 2022
The Nobel Prize in Chemistry is given by the Royal Swedish Academy of Sciences, Stockholm, Sweden.
The Nobel Prize in Chemistry 2022 was given to Carolyn R. Bertozzi, Morten Meldal, and K. Barry Sharpless "for the invention of click chemistry and bioorthogonal chemistry".

Sharpless and Meldal have built the framework for a functional kind of chemistry in which molecular building pieces snap together swiftly and effectively. Bertozzi has taken click chemistry to a new degree and begun exploiting it in live beings.

Their functional chemistry works wonders
Sometimes simple solutions are the best. Barry Sharpless and Morten Meldal are

awarded the Nobel Prize in Chemistry 2022 because they moved chemistry into the age of functionalism and built the foundations of click chemistry.

They share the award with Carolyn Bertozzi, who took click chemistry to a new level and started utilizing it to map cells. Her bioorthogonal reactions are now helping to better focus cancer therapies, among many other uses.

Nobel Prize in Physiology or Medicine 2022

The Nobel Prize in Physiology or Medicine is given by the Nobel Assembly at Karolinska Institutet, Stockholm, Sweden.

The Nobel Assembly at the Karolinska Institutet has agreed to present the 2022 Nobel Prize in Physiology or Medicine to Svante Pääbo "for his findings relating the genomes of extinct hominins and human evolution".

Through his pioneering study, Svante Pääbo did something virtually impossible: sequencing the genome of the Neanderthal, an extinct relative of present-day humans. He also made the spectacular finding of a previously undiscovered hominid, Denisova. Importantly, Pääbo also showed that gene transmission had happened from these now-extinct hominins to Homo sapiens during the journey out of Africa some 70,000 years ago.

This historical flow of genes to present-day humans has physiological importance today, for example impacting how our immune system responds to illnesses.

The Nobel Prize in Literature 2022
The Nobel Prize in Literature is given by the Swedish Academy, Stockholm, Sweden.

The Nobel Prize in Literature for 2022 is given to the French novelist Annie Ernaux "for the boldness and clinical clarity with which she discovers the origins, estrangements and communal restrictions of human memory".

In her work, Ernaux constantly and from diverse viewpoints, investigate a life characterized by major differences regarding gender, language, and class. Her route to literature was lengthy and tough.

The Nobel Peace Prize 2022

The 2022 Nobel Peace Prize is given to human rights campaigner Ales Bialiatski from Belarus, the Russian human rights group Memorial, and the Ukrainian human rights organization Center for Civil Liberties.

The Nobel Peace Prize laureates represent civil society in their nations. They have for many years advocated the right to question authority and preserve the basic rights of people. They have made an extraordinary effort to record war crimes, human rights violations, and the misuse of authority. Together they highlight the value of civil society for peace and democracy.

The economic sciences laureates 2022 (The Sveriges Riksbank Prize in Economic Sciences in Memory of Alfred Nobel).

The prize in economic sciences is granted by the Royal Swedish Academy of Sciences, Stockholm, Sweden, according to the same standards as for the Nobel Prizes that have been presented since 1901.

The Sveriges Riksbank Prize in Economic Sciences in Memory of Alfred Nobel 2022 was given to Ben S. Bernanke, Douglas W. Diamond, and Philip H. Dybvig "for research on banks and financial crises".
Ben Bernanke, Douglas Diamond, and Philip Dybvig, have substantially advanced our knowledge of the function of banks in the economy, especially during financial crises. A significant result of their study is why preventing bank crises is critical.

The Great Depression of the 1930s immobilized the world's economy for several years and had massive social ramifications. However, we have handled the following financial crises better due to study findings from this year's laureates. They have highlighted the necessity of averting massive bank crises.

Chapter 6: Nobel prize facts

The Nobel Prize is considered as the top honor in science. However, how much do we know about these awards?
Discover new and interesting things with these Nobel Prizes facts.

• Alfred Noble sponsored the Nobel Prize using explosives.
Alfred Nobel (1833-1896) was a Swedish scientist, engineer, and inventor. He was mainly interested in the firearms sector, with his most noteworthy creation being dynamite. This caused him to be called a trader of death. To fight against that legacy, Alfred ordered the Nobel Foundation formed in his will, with over 90% of his money going to the foundation. In that sense, he hoped that his legacy would amount to more than creating and selling guns.

• Separate academic institutions award the Nobel Prizes.
The Nobel Foundation only nominates and determines who wins the prizes for the year. Representatives of numerous academic institutions give the awards, as stipulated by Alfred Nobel in his will.
The Royal Swedish Academy of Sciences grants the Nobel Prizes for Physics and Chemistry.
The Karolinska Institute gives the Nobel Prize for Medicine.
The Norwegian Nobel Committee grants the Nobel Prize for Peace. Meanwhile, the Swedish Academy awards the Nobel Prize for Literature. Upon its inception in 1968, the Swedish Academy of Sciences grants the Nobel Prize for Economics.

• If a Nobel Prize laureate dies before being honored, the prize will be presented to their family.
Nobel Prizes aren't given posthumously. However, if the laureate dies after they're

pronounced the winner for that year, the prize is not withheld. Instead, it's handed to their family in their name.

• The Nobel Prize medals are composed of gold.
The Nobel Prizes come in 23-carat gold for those created before 1980. From 1980 until the present, the Nobel Prizes come in 18-carat green gold with 24-carat gold plating. The Swedish Mint created the medals until 2010, followed by the Mint of Norway which manufactures the medals until present.

• Alfred Nobel's visage is on the medal's reverse side, while the laureate's name is along the reverse side's rim. As for the front of the medal, the design varies on what the award is for. However, the front's rim is uniform with the phrase "Inventas vitam iuvat excoluisse per artes". Translated from its original Latin into English, it means "It is good to have enhanced life via discovering arts".

• The medals for physics and chemistry have two figures imprinted on the front.
The first figure for the Physics and Chemistry award is the goddess Isis, clutching a cornucopia in one hand, both signifying nature. The second figure is a lady symbolizing science, lowering Isis' veil and showing her face. This depicts intellect exploring the mysteries of nature.

• The medal for the Nobel Prize in Medicine includes a mother and a kid imprinted on the front.
The Nobel Prize in Medicine displays a lady sitting with an open book on her lap, signifying medicine. A girl stands next to her, with medication drawing water from a spring to give to the girl.

• The Nobel Prize for Literature medal evokes Greco-Roman mythology.

The award for the Nobel Prize in Literature portrays a young man signifying the artist, seated beneath a laurel tree. In Greco-Roman mythology, the laurel tree was sacred to Apollo, the god of arts and music. The medal also portrays one of the muses aiding the artist in his work.

• The Nobel Peace Prize features three men imprinted on the front.
The Nobel Peace Prize award portrays three men with their arms over each other's shoulders as a sign of fraternity. The writing along the front side's rim is likewise distinct from other medals. It says "Pro pace et fraternitate gentium", which means "For the peace and fraternity of men" when translated from its original Latin into English.

• The medal for Nobel Prize in Economic Sciences bears a star as its emblem.
This star symbolizes the Swedish Academy of Sciences, which awards the Nobel Prizes

for physics, chemistry, and economics. The inscription along the front side's rim is in Swedish, saying "Sveriges Riksbank till Alfred Nobels Minne 1968". The Swedish inscription translates to "The Sveriges Riksbank, in remembrance of Alfred Nobel, 1968."

This mentions the inception of the Nobel Prize for Economics in 1968, at the sponsorship of the Sveriges Riksbank.

• Nobel Prize laureates get their diplomas from the King of Sweden.

In addition to a medal and financial prize, Nobel Prize laureates get certificates for their achievement. The design for each diploma is unique to the laureate receiving it. Aside from the peace prize, all laureates have their diplomas given to them by the King of Sweden.

• The 1973 Nobel Prize for Peace is one of the most contentious.

In 1973, Henry Kissinger and Lê Đức Thọ earned a Nobel Peace Prize for negotiating a truce in the Vietnam War. Two members of the Norwegian Nobel Committee resigned in protest, since violence persisted in Vietnam despite the truce.

Critics condemned Kissinger as aggravating the situation as US President Nixon's National Security Adviser. In contrast, Lê Đức Thọ declined the award, on the premise that there was still no peace in Vietnam.

- Palestinian leader Yasser Arafat's Nobel Prize for Peace in 1994 also aroused criticism.

While it's disputed whether the Palestinian Liberation Group (PLO) is a terrorist organization, one of its allies is Hamas, recognized globally as a terrorist organization. This led prompted one member of the Norwegian Nobel Committee to quit in protest, at giving the peace prize to someone he regarded as a terrorist commander.

- The innovator of lobotomy got a Nobel Prize for Medicine.

As significant as the Nobel Prize is, it hasn't always recognized the greatest individuals. Antonio Moniz created the technique of lobotomy, a surgical procedure where surgeons cut off sections of the brain to cure mental disorders.

Now, lobotomy is done only as a last resort therapy. Today's lobotomy is also lot more polished than the formerly harsh approach. In the 1950s, tens of thousands of patients had lobotomy and suffered irreparable brain damage as a consequence.

- Mahatma Gandhi got a nomination for the Nobel Prize for Peace multiple times.

The first three occasions were from 1937 to 1939. The fourth time occurred in 1947, and lastly in 1948. After his killing, members of the Nobel Committee expressed sorrow that Gandhi never got the peace award.

- Several world-famous authors never got a Nobel Prize despite their accomplishments.
We're sure you've seen the Lord of the Rings movie series. You may even have read the books before or after the movie. Either way, we're sure you know how renowned J.R.R. Tolkien is. Strange as it appears, he never earned the literary award despite his tremendous achievements.

- Other renowned writers who never got the literary award include Leo Tolstoy, Emile Zola, and Mark Twain among others.

- The Nobel Prize encourages discoveries over innovations.
It's not that innovators aren't granted Nobel Prizes, but Nobel Prizes traditionally go to scientists first and inventors second. Statistics reveal that for the physics prize alone, 77% go to scientists, while just 23% is granted to innovators.

- Marie Curie is one of 4 persons to get 2 Nobel Prizes.

She initially got the physics prize in 1903, in appreciation of her work on radioactivity. She then got the chemistry prize in 1911 for isolating the radioactive element radium. So far, she is the only scientist to earn 2 Nobel Prize prizes.

- A Russian millionaire once purchased a Nobel Prize model.

In 2014, Alisher Usmanov acquired Dr. James Watson's medal at an auction in New York. Dr. Watson conducted the auction to generate money for cancer research, with Usmanov winning the model for $4.1 million.

To Dr. Watson's amazement, Usmanov returned the medal after the auction, claiming he deserved to retain the medal, and that the $4.1 million should continue to go to cancer research.

- Nobel Prize candidates don't get notified.

Up to 3000 persons every year are nominated for a Nobel Prize. However, they would only know their status if they win the award. Other than that, they may wait 25 years for the announcement of the current candidates.

• Anyone may be nominated for the Nobel Prize.
Adolf Hitler, Benito Mussolini, and Joseph Stalin were all Nobel Prize candidates in their time, even if they never knew it. None of them ever received the award, and it's no wonder they didn't.

www.ingramcontent.com/pod-product-compliance
Lightning Source LLC
Chambersburg PA
CBHW050315220526
45465CB00005B/2003